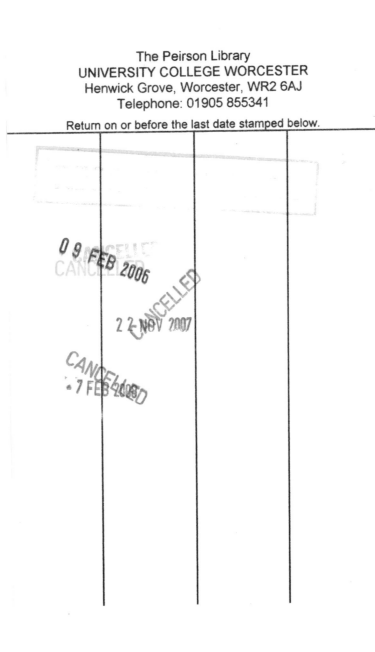

Metamorphic Rocks

Melissa Stewart

Heinemann
LIBRARY

H www.heinemann.co.uk/library
Visit our website to find out more information about Heinemann Library books.

To order:
☎ Phone 44 (0) 1865 888066
🖹 Send a fax to 44 (0) 1865 314091
💻 Visit the Heinemann Bookshop at www.heinemann.co.uk/library to browse our catalogue and order online.

First published in Great Britain by Heinemann Library, Halley Court, Jordan Hill, Oxford OX2 8EJ
a division of Reed Educational and Professional Publishing Ltd. Heinemann is a registered trademark
of Reed Educational and Professional Publishing Ltd.

OXFORD MELBOURNE AUCKLAND JOHANNESBURG BLANTYRE
GABORONE IBADAN PORTSMOUTH (NH) USA CHICAGO

© Reed Educational and Professional Publishing Ltd 2002
The moral right of the proprietor has been asserted.

Produced for Heinemann Library by Editorial Directions
Designed by Ox and Company
Originated by Ambassador Litho Ltd
Printed in Hong Kong

ISBN 0 431 14373 0
06 05 04 03 02
10 9 8 7 6 5 4 3 2

British Library Cataloguing in Publication Data
Stewart, Melissa
 Metamorphic rocks. – (Rocks and minerals)
 1. Rocks, metamorphic – Juvenile literature
 I. Title
 552.4

Acknowledgements
The Publishers would like to thank the following for permission to reproduce photographs:

Photographs ©: Cover background, H.H. Thomas/Unicorn Stock Photos; cover foreground, Martin Miller/Visuals Unlimited,
Inc.; p. 4, Cameramann International, Ltd.; p. 5, Grace Davies Photography; p. 7, John Springer/Bettmann/Corbis; p. 10 top,
Tom Bean; p. 10 bottom, Gerald & Buff Corsi/Visuals Unlimited, Inc.; p. 11, M. Long/Visuals Unlimited, Inc.; p. 12, Sylvester
Allred/Visuals Unlimited, Inc.; p.13, A.J. Copley/Visuals Unlimited, Inc.; p. 14, Keystone/The Image Works; p. 15, Joseph L.
Fontenot/Visuals Unlimited, Inc.; p. 17, Corbis; p. 18, Tom Bean; p. 19, Roger Ressmeyer/Corbis; p. 20, Maurice
Nimmo/Frank Lane Photo Agency/Corbis; p. 21, Townsend P. Dickinson/The Image Works; p. 22, Grace Davies Photography;
p. 23, Adam Tanner/The Image Works; p. 24 top, Doug Sokell/Visuals Unlimited, Inc.; p. 24 bottom, Gary Milburn/Tom
Stack & Associates; p. 26, Glenn Oliver/Visuals Unlimited, Inc.; p. 27, B. Daemmrich/The Image Works; p. 28, Grace Davies
Photography; p. 29, Mark E. Gibson/Visuals Unlimited, Inc.

Every effort has been made to contact copyright holders of any material reproduced in this book. Any omissions will be
rectified in subsequent printings if notice is given to the Publishers.

Our thanks to Alan Timms and Martin Lawrence of the Natural History Museum, London for their assistance in the
preparation of this edition.

Disclaimer

Contents

Any words appearing in the text in bold, **like this**, are explained in the Glossary.

What is a rock?

Rocks come in all shapes and sizes. Small rocks are often called 'pebbles'. Some people call medium-sized rocks 'stones'. A rock that is too large to pick up and carry around is sometimes called a 'boulder'. You see rocks every day, but do you ever stop to take a closer look? You can tell a lot about a rock by examining it for just a few minutes. Is it smooth or rough? Is it shiny or dull?

Taroko Gorge divides some of Taiwan's most impressive metamorphic rocks. The gorge was formed as the Limu River cut through 19 kilometres (12 miles) of marble and granite.

The next time you are travelling somewhere by car, look for places where workers have blasted through rock to build the road. Notice the rock's colours and patterns. Do you see layers that seem twisted or folded? If so, you are probably looking at metamorphic rock. Many mountains are made of metamorphic rock.

DID YOU KNOW?

Eclogite is a metamorphic rock. Some people call it Christmas tree rock because it is green with small, round pieces of a red mineral called garnet.

Metamorphic rock is one of three groups of rocks found in the world. The other two groups are **sedimentary rock** and **igneous rock**. We will learn more about how metamorphic rock forms later in this book. Each group of rocks forms in a different way, but they are all made of minerals.

Minerals

A mineral is a natural solid material. No matter where you find it, a mineral always has the same chemical makeup and the same structure. In other words, the **atoms** that mix together to form a specific mineral always arrange themselves in the same way. Most minerals have a **crystal** structure. Crystals usually have a regular shape and smooth, flat sides called faces.

Schist

Schist is a metamorphic rock that usually contains minerals of biotite or muscovite mica, with some quartz and an occasional garnet. The crystal structure of quartz is made up of silicon and oxygen atoms that are always arranged in the same way.

Schist is one kind of metamorphic rock. It forms at moderate pressure and moderate temperature. Schist contains mostly mica, some quartz and an occasional garnet, which is a **gemstone**.

SCIENCE IN ACTION

Geologists, who study the Earth, and petrologists, scientists who study rocks, can identify a rock by knowing where it came from and by looking at the **properties** of its minerals. They look at the colour, shininess and hardness of the minerals. They also study the size, shape and arrangement of the crystals.

Layers of the Earth

To understand metamorphic rocks, we need to know something about the structure of the Earth. Most of our planet is made of rock.

magma movement

inner core (solid)

outer core (liquid)

mantle

crust

The Earth is made up of layers. The thin outer layer of the Earth is the crust. The next layer, the mantle, is made of magma that is constantly moving. The core is made of an outer liquid layer and an inner solid core.

Soil

When you dig into the ground, you find soil. Soil is made of broken-up rock mixed with rotting plant and animal material.

The crust

If you dig deeper, you will eventually hit solid rock. The soil and the layer of hard rock beneath it make up the **crust**. The land and oceans are on the top part of the crust.

DID YOU KNOW?

Earth's crust can be anywhere from 5 to 69 kilometres (3 to 43 miles) thick. The thickest parts are below tall mountains of metamorphic rock. The thinnest parts are usually underneath the oceans.

Inside the Earth

Below the crust is a thick layer of **molten** magma. This is the hot, liquid rock that forms the Earth's **mantle**. The heat that keeps magma melted comes from the Earth's **core**. The outer core is made of metals that have melted to form a gooey liquid. The inner core is made of solid metals. The weight of these overlying layers presses down on the inner core. This pressure holds the **molecules** that make up the inner core so close together that they cannot turn into a liquid.

Hot magma near the core moves upwards towards the crust, while cooler magma moves down to take its place. Over millions of years, magma slowly circles through the mantle.

WHAT A TRIP!

Since ancient times, people have wondered what the inside of the Earth is like. In 1864, a French writer named Jules Verne wrote a science-fiction novel called *Journey to the Centre of the Earth*. In 1959, the book was made into a film (left). Both tell the story of four people who enter a **volcano** on Iceland and travel all the way to the Earth's fiery core. Along the way, they see a cave filled with enormous mushrooms and dinosaur-like monsters that live far below the Earth's surface. Of course, there aren't really mushrooms or giant reptiles deep underground.

Land on the move

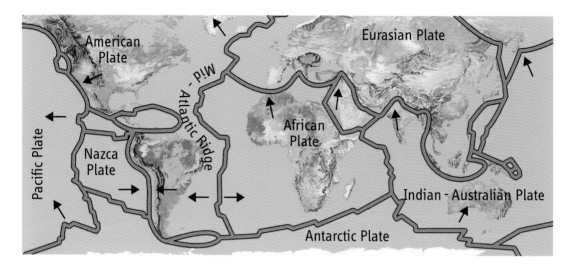

The Earth's surface is broken into many plates. The major plates are labelled on this diagram. The plates are moving constantly, though very slowly, in the direction of the arrows. The Mid-Atlantic Ridge is a rift formed by two plates moving apart.

DID YOU KNOW?

The scientific study of how the Earth's plates have moved over time is sometimes called tectonics. The word 'tectonics' comes from a Greek term meaning 'to put together or take apart'.

In some ways, the Earth's **crust** has a lot in common with the crust on top of a meat pie. The pie's crust rests on top of thick, steaming-hot gravy full of meat and vegetables. The Earth's crust rests on top of a sea of molten **magma**. Before you eat a meat pie, you cut its crust into pieces. The Earth's crust is broken into large pieces called **tectonic plates**.

Creating rifts

As magma circles through the **mantle**, the plates that make up Earth's crust move, too. In some parts of the world, the plates move apart and long cracks called **rifts** are left behind. When rifts form under the ocean, the material from the mantle rises to the surface and creates new land on each side of the rift.

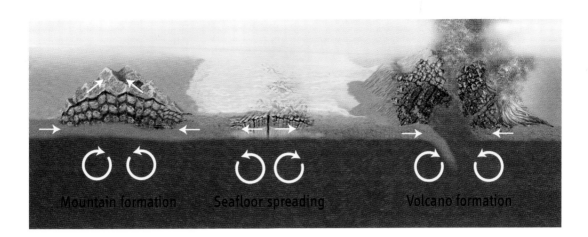

Mountain formation   Seafloor spreading   Volcano formation

This process is called **seafloor spreading**. When rifts form on land, earthquakes may shake the ground and **volcanoes** may erupt, spilling lava out over the land.

Making mountains

In other parts of the world, the plates bump into one another. Sometimes one plate slides over the other. Then the bottom plate moves down into the mantle, where it melts. When two plates crash and push against each other with great force, the land buckles and tall mountains form. When two plates scrape against each other, the result is a transform fault, such as the San Andreas Fault in California and the Dead Sea Fault in the Middle East. When enough pressure builds up along a fault, an earthquake occurs.

Mountains may form when two plates hit head-on. The seabed expands as magma rises through a rift. When one plate moves below another, magma may rise to the surface and escape through a volcano.

MOUNTAINS IN MOTION

Mount Everest, the tallest mountain in the world, is made of rock. It is part of the Himalaya Mountains. The mountains in this range get a little taller every year as the Indian-Australian Plate crashes into the Eurasian Plate.

Three groups of rocks

Limestone is one kind of sedimentary rock. It is made mainly of calcite.

As we have seen, the Earth has three groups of rocks – **sedimentary**, **igneous**, and metamorphic. Each group of rocks forms in a different way.

Igneous rock

Igneous rocks form when **magma** from the Earth's **mantle** cools and forms **crystals**. Sometimes the magma forces its way to the Earth's surface and spills onto the land as lava. This kind of igneous rock cools quickly and has very small crystals.

In other cases, pools of magma become trapped at some depth and cool slowly over thousands of years. Some of the largest and most beautiful crystals in the world were formed in this way. Granite, gabbro, basalt and obsidian are examples of igneous rock.

Obsidian is a dark, volcanic glass. It forms when lava cools very quickly.

Sedimentary rock

Sedimentary rocks form as layers of mud, clay, sand and other materials build up over time. The weight of the materials at the top of the pile presses down on the materials below. All that pressure cements the materials together to form hard rock. If you look closely at sedimentary rock, you may be able to see its layers. Limestone, sandstone, shale and conglomerate are examples of sedimentary rock.

Marble is a metamorphic rock that forms when limestone is squeezed and recrystallized. If the limestone was pure, the marble will be white. If the limestone contained impurities, the marble may be black, green, red or yellowish brown.

Metamorphic rock

Metamorphic rocks form when heat or pressure changes the **minerals** within igneous rock, sedimentary rock or another metamorphic rock. This often happens when the Earth's **tectonic plates** collide. Marble, slate, gneiss, hornfels, phyllite, migmatite, serpentinite, quartzite and schist are examples of metamorphic rock.

IMAGINE THAT!

Would you like a dog or cat, but your parents say, 'Not a chance'? In the 1970s, kids who wanted an easy-to-care-for companion found a new idea – the pet rock. For a few years, pet rocks were sold in shops around the world. Whether they were igneous, sedimentary or metamorphic, these unusual 'pets' had many advantages. They didn't need to be fed or walked. They were also small enough to fit in a person's pocket, and they were very well behaved!

How does metamorphic rock form?

You can clearly see the folds in the metamorphic rock that makes up the Canadian Rockies. The Rocky Mountains formed about 65 million years ago.

DID YOU KNOW?

As **sedimentary** and igneous rocks transform into metamorphic rock, **atoms** from one mineral sometimes sneak into a different mineral's crystal structure. When this happens, beautiful new crystals of sapphire, ruby, peridot, garnet, emerald and other prized **gemstones** may form.

The extreme pressure and temperature conditions that create metamorphic rocks cause the **minerals** in the rocks to change into different minerals. Most metamorphic rocks are made when sedimentary rock is heated, folded, and squeezed during mountain building. This process is clearly shown in the exposed layers of rock that form the peaks of the Canadian Rocky Mountains in British Columbia (see pages 14–15 for more details).

Hot magma

Other metamorphic rocks form when rocks are heated by streams of **magma** that spurt up into the crust. The sizzling-hot magma bakes the surrounding rock, turning it into metamorphic rock. Meanwhile, the magma itself eventually cools and forms **igneous rock**. Examples of this

HOW HEAT AND PRESSURE CHANGE ROCK

ORIGINAL ROCK	NEW METAMORPHIC ROCK
Basalt (igneous)	Serpentinite
Granite (igneous)	Gneiss
Limestone (sedimentary)	Marble
Sandstone (sedimentary)	Quartzite
Shale (sedimentary)	Slate
Slate (metamorphic)	Phyllite
Phyllite (metamorphic)	Schist
Schist (metamorphic)	Gneiss

process are more common below the Earth's surface. For this type of metamorphic rock to be seen, all the material above and around the area must be **eroded** over millions of years.

Pressure and heat

Because the minerals that make up metamorphic rock have had to endure tremendous heat and pressure, the resulting rock is hard and tough. Gneiss is especially durable because it forms when other kinds of metamorphic rock are altered. This rock usually has large, interlocking **crystals**. The colour of a gneiss specimen depends on the minerals it contains.

Gneiss is one of the toughest kinds of rock in the world. Because it forms deep underground, it can be seen only in uplifted mountain ranges or in severely eroded areas.

A look at mountains

The Alps in France and Switzerland formed more than two million years ago as the African Plate collided with the Eurasian Plate.

The Himalayas in Asia, the Alps and the Caledonian Mountains in Europe, and the Appalachian Mountains and Rocky Mountains in North America all contain metamorphic rock.

Most mountains form when the Earth's **tectonic plates** collide. Over long periods of time, ocean floor sediments are squeezed between colliding plates. The sediments are heated and folded into metamorphic mountain ranges.

Block mountains

Block mountains form when plates are pulled apart causing continents to crack into separate

DID YOU KNOW?

Tall mountain ranges sometimes affect weather by blocking cloud movements. As a result, it is often very wet on one side of a mountain range and very dry on the other side. Think about this as you look at the locations of the Earth's deserts.

blocks. Some blocks drop down to form valleys while others are left at higher levels to form mountains. This is what happened in Death Valley in the American states of California and Nevada.

Volcanoes

Other mountains are igneous. For example, some mountains are made of lava – **magma** that spills onto the Earth's surface through a **volcano**. As the lava cools, it hardens and forms mountain-sized piles. Some of the most spectacular volcanic mountains in the world include Mount Kilimanjaro in Tanzania, Africa, Mount Etna in Sicily, Italy and Mauna Loa in Hawaii, in the North Pacific Ocean.

A CLUE TO THE PAST

Scientists have discovered that the Appalachian Mountains (above) in the United States are part of the same ancient mountain range as the Caledonian Mountains that run through Scotland and Norway. How is this possible? Millions of years ago, North America and Europe were part of the same huge continent. Today, the American Plate is moving away from the Eurasian Plate, and the Atlantic Ocean is getting a little wider every year.

Too hot to touch

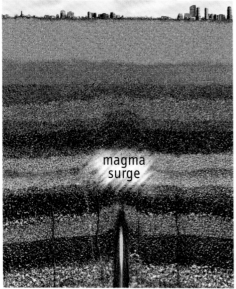

magma surge

In some parts of the world, magma intrudes through the crust. The rock that comes into contact with the hot magma is transformed into metamorphic rock.

When the Earth's **tectonic plates** collide, the force affects large areas of rock. That is why metamorphic mountain chains are often thousands of miles long. Some smaller areas of metamorphic rock form in a different way. Sometimes molten **magma** intrudes into the Earth's **crust**.

Superheating

Occasionally these **intrusions** get all the way to the Earth's surface and spill onto the land through **volcanoes**. In many cases, however, the magma is not able to travel all the way through the crust. The trapped magma is often over 1200°C and superheats the surrounding rock. As a result, the **minerals** in the rocks that touch the magma can change quite a bit. Further away, the scorching magma has less effect.

DID YOU KNOW?

Hornfels is a metamorphic rock that forms when shale, a sedimentary rock, comes into contact with magma. As the magma superheats the rock, the particles that make up shale are recrystallized to form hornfels.

As the magma cools, it forms a column of **igneous rock**. At the same time, nearby rock is transformed into metamorphic rock. Split Mountain in California, USA is a perfect example of this process.

16

Examining Split Mountain

When you look at some parts of the Split Mountain formation, you can see a dark layer of metamorphic rock on top of a lighter layer of igneous rock. The igneous rock was once magma. It heated the **sedimentary rock** above and turned it into metamorphic rock. Later, the entire area was lifted up and became part of the Earth's surface.

Split Mountain in California has distinct layers of metamorphic and igneous rock. This provides clear evidence that magma once intruded this area.

A shocking experience

Most metamorphic rocks form slowly over millions of years. The tremendous heat and pressure inside the Earth need time to affect **minerals** in ways that transform them into metamorphic rocks. But there is a way to make metamorphic rocks at lightning speed.

A smooth, glassy line of fulgurite runs through this rock. This tells scientists that lightning once struck the rock.

Shock metamorphism

When a bolt of lightning strikes sand, the grains may fuse together and form a solid metamorphic rock called fulgurite. Scientists call this process shock metamorphism. Fulgurite is very fragile, so it is unusual to find big pieces of this rock. The world's longest fulgurite sample was found by researchers at the University of Florida in the United States in 1996. This shocked strip is 5.2 metres long.

Meteorite craters

Shock metamorphism can also happen when a **meteorite** strikes the Earth's surface. As a space rock falls through the Earth's **atmosphere**, particles in the air rub against it and create a force called **friction**. Friction generates a lot of heat. When the object crashes into the ground, it can transfer all its heat energy to whatever it hits. As the force and energy strike our planet's surface, minerals in the rocky layers close to the surface can instantly rearrange themselves to form metamorphic rock.

Meteorite craters lined with metamorphic rock have been discovered in many parts of the world, including Australia, Mexico and the USA. Other meteorite craters are hidden under the oceans. Giant meteorite craters can also be seen on the surfaces of Mercury, Mars and many of the moons in our solar system.

WHAT A DISCOVERY!

In the 1950s, Eugene Shoemaker began to study Meteor Crater near Flagstaff, Arizona, USA. Shoemaker was a geologist – a scientist who studies rocks to try to understand how the Earth formed and how it has changed over time. By studying the crater, Shoemaker discovered that the area was once made of **sedimentary rock**. About 25,000 years ago, a tremendous force had suddenly transformed rock at the surface of the crater into metamorphic rock. This proved that a meteorite had created Meteor Crater. Although Meteor Crater is about 1220 metres across, the space rock that created the giant hole was only about 30 metres wide.

How do people use metamorphic rock?

DID YOU KNOW?

In the past, most gravestones were made of slate, which is a kind of metamorphic rock. Today, nearly all gravestones are made of granite, a kind of **igneous rock**.

Rock is one of the Earth's most valuable **natural resources**. Some types of rock contain deposits of oil and natural gas. Important **minerals**, metals and precious **gemstones** are found in rock, too.

Rhodochrosite, a mineral found in metamorphic rock, contains the metal manganese, which is added to iron during the steel-making process.

The metamorphic rock schist is mined because it sometimes contains large garnet **crystals**. Garnet is usually blood-red, but it may also be yellow, light green or colourless. This gem has been prized for centuries and is frequently used to make jewellery.

Important minerals

The minerals chromite and rhodochrosite are sometimes found in metamorphic rock. Chromite contains the metal chromium – an important ingredient in stainless steel. Chromium makes stainless steel hard and prevents it from rusting. Rhodochrosite is used to make batteries and to purify drinking water.

Talc and graphite are minerals that form only in metamorphic rock. Because talc is readily available, soft and heat resistant, it is used to make baby powder, paper, paints, soap, fireproof roofing, linoleum, electrical insulation and pottery. Graphite, another soft mineral, is used as the 'lead' in pencils.

Serpentine

Serpentinite is a metamorphic rock that looks similar to green marble, and the two rocks are often confused. But serpentinite and marble contain different minerals. Serpentinite is made of the mineral serpentine. Marble contains calcite or dolomite and small amounts of other minerals. Like marble, serpentinite is soft enough to carve and polish. Many artists create beautiful sculptures and jewellery from serpentinite.

Quartzite is a very hard rock made of tightly packed recrystallized sand. It is sometimes used to build roads. It is also the perfect material for making millstones used to grind wheat into flour in flour mills.

Slate splits easily into flat pieces. It is sometimes used to make roof tiles, like those seen here.

Magnificent marble

By far the most widely used metamorphic rock is marble. This important metamorphic rock forms when layers of limestone are exposed to tremendous heat and pressure. Large regions of natural marble can be seen in the Alps and on the Greek islands in the Mediterranean.

Different colours

Marble is beautiful and comes in many colours, including pure white, black, green, red and yellowish brown. It is also easy to cut and polish. The Lincoln Memorial in Washington, D.C. and parts of the Parthenon in Athens, Greece, are made of marble.

The marble-built Taj Mahal houses the tombs of Mumtaz Mahal and her husband, Shah Jahan. The main building took eleven years to build, and the smaller buildings and gardens took another ten years to complete.

DID YOU KNOW?

Most of the world's caves are carved out of **sedimentary rock**. Oregon Caves National Monument in the USA is one of the few sites where water has slowly dissolved tough metamorphic rock to create long, winding passageways full of beautiful rocky formations.

Taj Mahal

In the early 1600s, Mumtaz Mahal, the beautiful young wife of the Indian emperor, Shah Jahan, died unexpectedly. The emperor decided to build a stunning structure in memory of his beloved wife. The Taj Mahal is made of huge slabs of pure white marble that came from a quarry located more than 483 kilometres (300 miles) away from the building site at Agra in India. The heavy rock had to be transported in special carts pulled by oxen and elephants.

Marble as art

People also use marble to make furniture, staircases, floor tiles, kitchen and bathroom work surfaces and more. It can even be carved with a hammer and chisel to create beautiful sculptures. The famous Italian painter, sculptor and architect Michelangelo (1475–1564) carved many incredible marble sculptures. All the rock Michelangelo used came from the Carrara quarry in Tuscany in Italy.

Marble is still being removed from this quarry near Carrara in Italy. The rock was created as the Apennine Mountains were lifted up millions of years ago.

The rock cycle

Rocks are always changing. As the Earth's **tectonic plates** ram into one another, rock deep below the surface is compressed and folded, lifting the land to create mighty mountains. In the process, metamorphic rock is formed.

Weathering caused the fracture in this boulder. Heat and cold cause rocks to expand and contract, creating cracks that are deepened by water and ice.

Weathering

Other metamorphic rock is exposed when water, wind or ice attacks the overlying rock. Rocks can also be broken down by plant roots, chemicals produced by tiny creatures that live in the soil and repeated freezing and thawing. This is called **weathering**.

Have you ever seen a boulder that looked as though it had mysteriously split in half? The split was probably the result of weathering.

DID YOU KNOW?

Metamorphic rock is very hard, which can slow down erosion and weathering. Nevertheless, marble sculptures around the world are threatened by **acid rain**. The chemicals in the rain slowly wear away the marble.

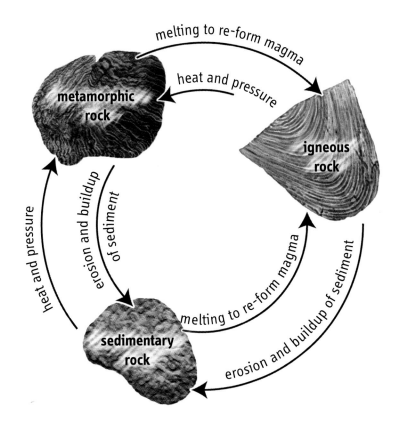

The rock we see on the Earth today has not always been here. Rock forms and breaks down in a never-ending cycle.

As rock breaks down and wears away, the sediments are picked up by glaciers or wind or rivers and streams. Eventually, these sediments may travel all the way to the ocean. Over time, layers of sediment build up and form **sedimentary rock**.

As the Earth's plates move, some of the sedimentary rock is heated so much that it first changes to metamorphic rock and then melts completely to become **magma**. Eventually, some of that magma will cool to form **igneous rock**. Other areas of the sedimentary rock will be pressed and twisted. Under this tremendous pressure, they will form more metamorphic rock.

MOUNTAINS DON'T LAST FOREVER

The Rockies, the Andes and the Himalayas all began to form between 60 and 70 million years ago. This makes them fairly young, with high peaks and sharp ridges. The Hercynian Mountains of Devon and Cornwall in England are much older. Over time, much of the material that once made up these mountains has been worn away by wind and water. They are now so eroded away that all you can see is their granite core.

Identifying metamorphic rocks

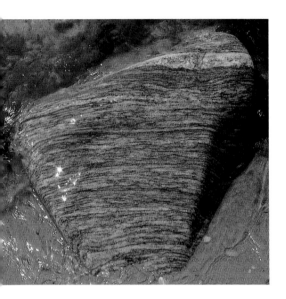

The banding in this gneiss sample makes it easy to identify it as a metamorphic rock. Have you ever seen rock with similar light and dark layers as you drove along a road cutting?

DID YOU KNOW?

Scientists have found quartzite in Australia that is 3.5 billion years old. It may be the oldest metamorphic rock on the Earth's surface.

Now that you've learned about metamorphic rocks, you might want to search for some. How can you tell a metamorphic rock from other kinds of rocks? You can look for some telltale signs.

Bands in the rock

First, think about how metamorphic rock forms. Tremendous heat and pressure cause the **minerals** in other kinds of rock to be compressed, twisted and rearranged. As a result, the minerals in metamorphic rock often face in the same direction and form light and dark bands. These characteristics are especially easy to see in samples of schist and gneiss. Sometimes the bands in metamorphic rock look stretched out and are easy to break into sheets. That's why people make roof and floor tiles out of slate.

Location

Knowing where a rock comes from can also help you identify it. For example, if you find a rock near an ancient mountain chain, it may be a metamorphic rock. Metamorphic mountains often formed in places where large landmasses collided as the Earth's **tectonic plates** moved.

To find out even more about a rock, you can study its **crystals**. Their colour, shininess and hardness can help you identify the minerals in the rock. Petrologists also pay close attention to the size, shape and arrangement of the crystals in a rock. The crystals that make up metamorphic rocks are usually small.

Of course, one of the best ways to identify a rock is to study a field guide to rocks and minerals. These books show pictures of rocks and give detailed descriptions of them.

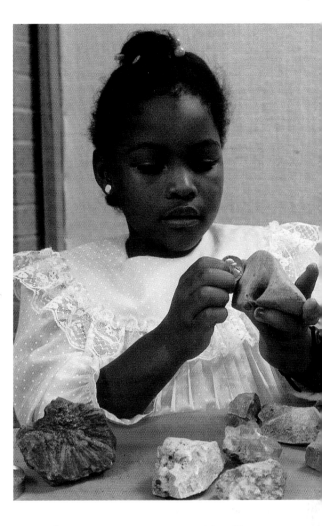

You can identify rocks using the same techniques as scientists. Begin by studying the characteristics of the minerals that make up the sample.

CLUES FROM THE PAST

Fossils are most common in **sedimentary rock**, but they also occur in metamorphic rock that formed from sedimentary rock. The fossils in metamorphic rock often show signs of what the rock has endured. The ancient remains may be crushed, broken, squeezed or stretched out.

Rock collecting

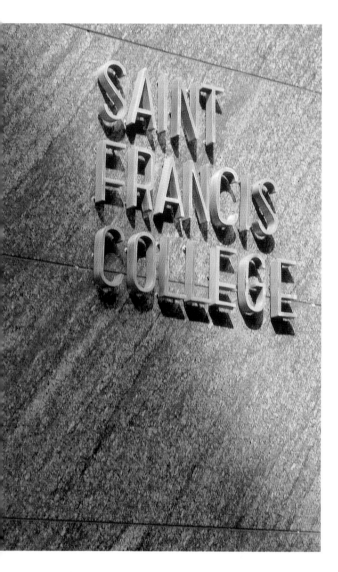

The exterior of this building is made of gneiss. The banding pattern gives you a clue that this is a metamorphic rock.

Now that you know how to spot metamorphic rocks, you can begin searching for them. Look for marble in buildings and slate on roofs and floors. You might also want to plan a rock hunting trip. There are many rocks to search out in the woods, in fields or maybe even in your own garden. Before you head out to an unfamiliar place, however, you will need to gather a few pieces of equipment and learn a few basic rules.

Be systematic!

Once you have identified the rocks, it's a good idea to create a system for labelling, organizing and storing them. Then you will always be able to find a specific sample later. You can arrange your specimens any way you like – by colour, by **crystal** shape, by collection site or even alphabetically. As your collection grows, being organized will become more and more important.

This impressive rock display belongs to a university collection in the United States. Each rock has been carefully labelled by a scientist.

WHAT YOU NEED TO KNOW

- Never go rock hunting alone. Go with a group that includes a qualified adult.
- Know how to use a map and compass.
- Always get a landowner's permission before walking on private property. If you find interesting rocks, ask the owner if you can remove them.
- Before removing samples from public land, make sure rock collecting is allowed. Many natural rock formations are protected by law.
- Respect nature. Do not hammer out samples. Do not disturb living things and do not leave litter.

WHAT YOU NEED

- Strong boots or wellies
- A map and compass
- A small paintbrush to remove dirt and extra rock chips from samples
- A camera to take photographs of rock formations
- A hand lens to get a close-up look at **minerals**
- A notebook for recording when and where you find each rock
- A spotter's guide to rocks and minerals.

Glossary

acid rain rain that is polluted with acid in the atmosphere and that damages the environment

atmosphere layer of air and other gases that surrounds the Earth

atom smallest unit of an element that has all of the properties of the element

core centre of the Earth. The inner core is solid and the outer core is liquid.

crust outer layer of the Earth

crystal repeating structure within most minerals

erode to slowly wear away rock over time by the action of wind, water or glaciers

fossil remains or evidence of ancient life

friction force that resists motion between two objects or surfaces. If there is motion and friction is created, energy is converted to heat.

gemstone beautiful mineral that may be worn as jewellery

igneous rock rock that forms when magma from the Earth's mantle cools and hardens

intrusion when magma forces itself into cracks in the Earth's crust and heats the surrounding rock

magma hot, liquid rock that makes up the Earth's mantle. When magma spills out onto the Earth's surface, it is called lava.

mantle layer of the Earth between the crust and outer core. It is made of rock in its liquid form, known as magma.

meteorite chunk of rock from space that hits the Earth or another object

mineral natural solid material with a specific chemical makeup and structure

molecule smallest unit of a substance, made up of one or more atoms

natural resource natural material that humans use to make important products

property characteristic that helps make identification possible

rift crack in the Earth's surface created when two plates move apart

seafloor spreading process that occurs when the Earth's plates move apart, creating a crack on the floor of the ocean

sedimentary rock rock that forms as layers of mud, clay and tiny rocks build up over time

sediment mud, clay or bits of rock picked up by rivers and dumped in the ocean

tectonic plate one of the large slabs of rock that make up the Earth's crust

transform fault crack that forms on the Earth's surface where two plates scrape against each other

volcano opening in the Earth's surface that extends into the mantle

weathering breaking down of rock by plant roots or by repeated freezing and thawing

Further information

BOOKS

The Kingfisher book of planet Earth, Martin Redfern, Kingfisher, 1999

The pebble in my pocket, Meredith Hopper, Frances Lincoln, 1997

The best book of fossils, rocks and minerals, Chris Pellant, Kingfisher, 2000

Tourists rock, fossil and mineral map of Great Britain, British Geological Survey, 2000

ORGANIZATIONS

British Geological Survey
www.bgs.ac.uk
Kingsley Dunham Centre, Keyworth,
Nottingham, NG12 5GG UK

Rockwatch
www.geologist.demon.co.uk/rockwatch/
The Geologists' Association
Burlington House, Piccadilly,
London, W1V 9AG UK

The Natural History Museum
www.nhm.ac.uk
Cromwell Road,
London, SW7 5BD UK

The Geological Society of Australia
www.gsa.org.au/home
Suite 706, 301 George Street
Sydney NSW
Australia 2000

Geological Survey of Canada
www.nrcan.gc.ca/gsc/
601 Booth Street
Ottawa, Ontario
KIA 0E8
Canada

US Geological Survey (USGS)
www.usgs.gov
507 National Center
12201 Sunrise Valley Drive
Reston, Virginia 22092
USA

Index

Titles in the *Rocks and Minerals* series include:

Hardback   0 431 14370 6

Hardback   0 431 14371 4

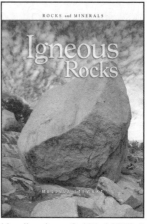

Hardback   0 431 14372 2

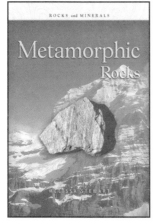

Hardback   0 431 14373 0

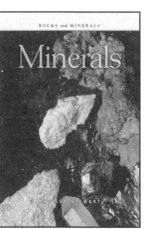

Hardback   0 431 14374 9

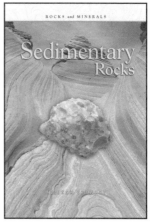

Hardback   0 431 14375 7

Hardback   0 431 14376 5

Find out about the other titles in this series on our website www.heinemann.co.uk/library